Animal Migrations

Bird Migration

by Jen Breach

FOCUS READERS®
BEACON

www.focusreaders.com

Focus Readers is distributed by North Star Editions:
sales@northstareditions.com | 888-417-0195

Produced for Focus Readers by Red Line Editorial.

Photographs ©: Shutterstock Images, cover, 1, 4, 7, 8, 11, 13, 14–15, 16, 18, 21, 22, 25, 27, 29

Library of Congress Cataloging-in-Publication Data
Names: Breach, Jen, author.
Title: Bird migration / by Jen Breach.
Description: Lake Elmo, MN : Focus Readers, [2024] | Series: Animal migrations | Includes index. | Audience: Grades 2-3
Identifiers: LCCN 2022056595 (print) | LCCN 2022056596 (ebook) | ISBN 9781637396056 (hardcover) | ISBN 9781637396629 (paperback) | ISBN 9781637397732 (pdf) | ISBN 9781637397190 (ebook)
Subjects: LCSH: Birds--Migration--Juvenile literature. | Migratory birds--Juvenile literature.
Classification: LCC QL698.9 .B74 2024 (print) | LCC QL698.9 (ebook) | DDC 598.156/8--dc23/eng/20221223
LC record available at https://lccn.loc.gov/2022056595
LC ebook record available at https://lccn.loc.gov/2022056596

Printed in the United States of America
Mankato, MN
082023

About the Author

In 2012, Jen Breach (they/them) migrated 10,371 miles (16,700 km) from Melbourne, Australia, to New York City. Jen has worked as a bagel-baker, a code-breaker, a ticket-taker, and a trouble-maker. They now work as a writer, the best job ever, in Philadelphia, Pennsylvania.

Table of Contents

Chapter 1

From Pole to Pole

It is a chilly March morning. An Arctic tern skims over the ocean. Its light body glides through the air. The tern swoops down and grabs a fish. It keeps flying as it eats. It has a long way to go.

Arctic terns eat small fish that they can easily carry away.

The tern is flying from Antarctica to the Arctic. It will arrive in April or May. Then it will make a nest and lay eggs. In August, the tern will fly back to Antarctica. That is where terns feed. They build up energy for the next **migration**.

An Arctic tern can fly more than 900,000 miles (1,450,000 km) in its lifetime. That's like going to the moon and back twice.

Arctic terns can live for more than 30 years.

Terns make this journey every year. Each time, they fly more than 30,000 miles (48,000 km). That is the longest migration of any animal.

Chapter 2

Why Migrate?

All birds need food to eat. They also need places to lay their eggs. Some birds eat and lay eggs in the same area. But for other birds, feeding areas and nesting areas are far apart. So, they must migrate.

More than 4,000 bird species migrate. That's nearly 40 percent of all birds.

A good nesting place must have the right weather. It can't be too hot or too cold. Otherwise, the eggs will not hatch. Also, the nesting place must be safe. It could be an area with few **predators**. Or it could be an area with good **camouflage.** That way, the baby birds have the best chance of survival.

Some **species** have long migrations. Examples include warblers and orioles. These songbirds fly between North

A nighthawk keeps its eggs safe by blending in with its environment.

America and Central America each year. Turtledoves and honey buzzards also have long migrations. They fly between Europe and Africa.

Some birds migrate in groups. Geese are one example. Geese teach migration routes to their young. They return to the same place year after year.

Other birds avoid large groups. They fly apart from one another. These birds migrate by **instinct**. Scientists aren't exactly sure how

Geese are high flyers. Some fly up to 5.5 miles (8.9 km) above sea level.

Flying in a V-shaped pattern helps geese save energy.

they find their way. The birds might use the sun and stars. Or they might use Earth's **magnetic field**.

ANIMAL SPOTLIGHT

Northern Wheatears

Northern wheatears are small songbirds. In the summer, they live in Alaska and Canada. That is where they lay eggs. When fall comes, they migrate to Africa. That is where they feed.

Some wheatears fly east when they migrate. They cross the Atlantic Ocean. After that, they fly south to Africa. However, other wheatears fly west when they migrate. They cross Asia. Then they fly across land to reach Africa. Both groups end up in the same area. But they take very different routes to get there.

Northern wheatears migrate thousands of miles each year.

Chapter 3

Medium and Short Migrations

Not all migrations cover long distances. Some birds have medium-distance migrations. They fly only a few hundred miles. This usually happens when food sources run low.

American tree sparrows migrate medium distances.

Cedar waxwings eat fruit, cones, and insects.

Some medium-migration birds are nomadic. That means they travel around a large area. They search for food. For instance, cedar waxwings forage for berries. These birds may go to different areas every year. It depends how much food is available. If there is plenty of food, they stay where they are.

Some birds don't migrate every year. For example, certain blue jays might fly north and south.

Not all migratory birds fly. Penguins migrate by swimming. Emus walk hundreds of miles.

But others might stay where they are. It often depends on the bird's age. Younger blue jays are more likely to migrate. Older blue jays are more likely to remain in one place.

Some birds stay in the same general area. But they migrate to different **elevations**. For example, yellow-eyed juncos remain in Mexico all year. They nest high in the mountains. That way, they are safer from predators. But they feed in lowlands. In these areas,

During winter, yellow-eyed juncos eat many seeds.

there is more food. The birds may migrate less than 1 mile (1.6 km). But the two **environments** are very different.

Chapter 4

Dangers of Migration

Birds migrate to safe nesting areas. This gives their babies a better chance of survival. But migration can be dangerous. Exhaustion is one risk. Some birds travel long distances without a place to rest.

Ruby-throated hummingbirds can cross large bodies of water without stopping.

For instance, some birds fly across oceans. Ruby-throated hummingbirds spend winters in Central America. To get there, they fly across the Gulf of Mexico. This trip takes a lot of energy. So, the hummingbirds double their weight before leaving.

Humans are the biggest danger for most birds. Migrating birds often rest in forests. The trees offer good protection from predators. Forests also have plenty of food.

Between 2010 and 2020, the world lost millions of acres of forest each year.

But humans are cutting down many forests. If birds cannot rest in these places, they could starve. They could also get eaten by predators.

Humans often clear land for new buildings. However, many birds die each year from hitting buildings.

Also, bright city lights can cause birds to lose their way. Pets are another major problem. Each year, cats kill billions of birds.

There are ways to help migratory birds. People can keep their cats indoors. They can turn off lights at night. People can also contact

Some people paint designs on their windows. That way, birds are less likely to crash into them.

Small stickers that reflect light can prevent birds from crashing into windows.

lawmakers. Lawmakers have the power to make big changes. They can protect forests where birds feed.

FOCUS ON
Bird Migration

Write your answers on a separate piece of paper.

1. Write a letter to a lawmaker explaining the dangers that migrating birds are facing.
2. Do you think humans are doing enough to protect migrating birds? Why or why not?
3. Which bird has the longest migration?
   - **A.** blue jay
   - **B.** cedar waxwing
   - **C.** Arctic tern
4. Why do some birds migrate instead of staying in their nesting areas?
   - **A.** There isn't enough food in the nesting area.
   - **B.** There are not enough predators in the nesting area.
   - **C.** There are too many eggs in the nesting area.

5. What does **forage** mean in this book?

*They search for food. For instance, cedar waxwings **forage** for berries.*

A. try to find something
B. fly quickly past something
C. hide from something

6. What does **exhaustion** mean in this book?

***Exhaustion** is one risk. Some birds travel long distances without a place to rest.*

A. traveling a short way
B. resting for a long time
C. getting very tired

Answer key on page 32.

Glossary

camouflage

Colors that make something difficult to see in the area around it.

elevations

Heights of land in relation to the sea.

environments

The natural surroundings of living things in particular places.

instinct

A behavior that is natural instead of learned.

magnetic field

The space around an object (such as a moon or planet) in which its magnetic force can be detected.

migration

The movement of a group of animals from one place to another.

predators

Animals that hunt other animals for food.

species

Groups of animals or plants that are alike and can breed with one another.

To Learn More

BOOKS

Abell, Tracy. *Birds*. Minneapolis: Abdo Publishing, 2021.

MacCarald, Clara. *Migrating to Survive*. Minneapolis: Abdo Publishing, 2023.

Perdew, Laura. *Investigating Light Pollution*. Mankato, MN: The Child's World, 2022.

NOTE TO EDUCATORS

Visit **www.focusreaders.com** to find lesson plans, activities, links, and other resources related to this title.

Index

Answer Key: 1. Answers will vary; **2.** Answers will vary; **3.** C; **4.** A; **5.** A; **6.** C